GRAYSCALE
Bouquet

COLORING BOOK
FOR ADULTS

Vol.1

COLOR TEST PAGE

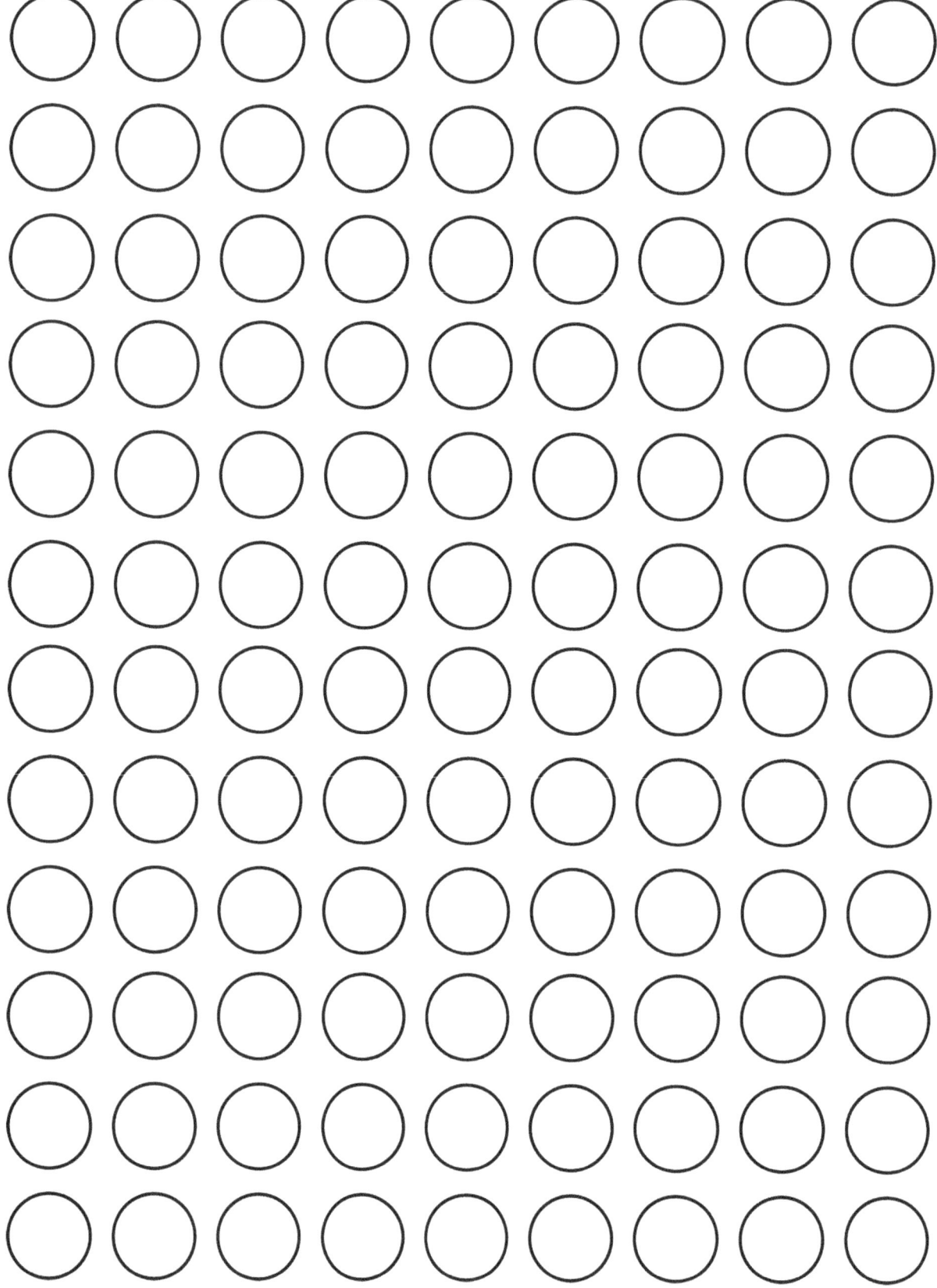

COLOR TEST PAGE

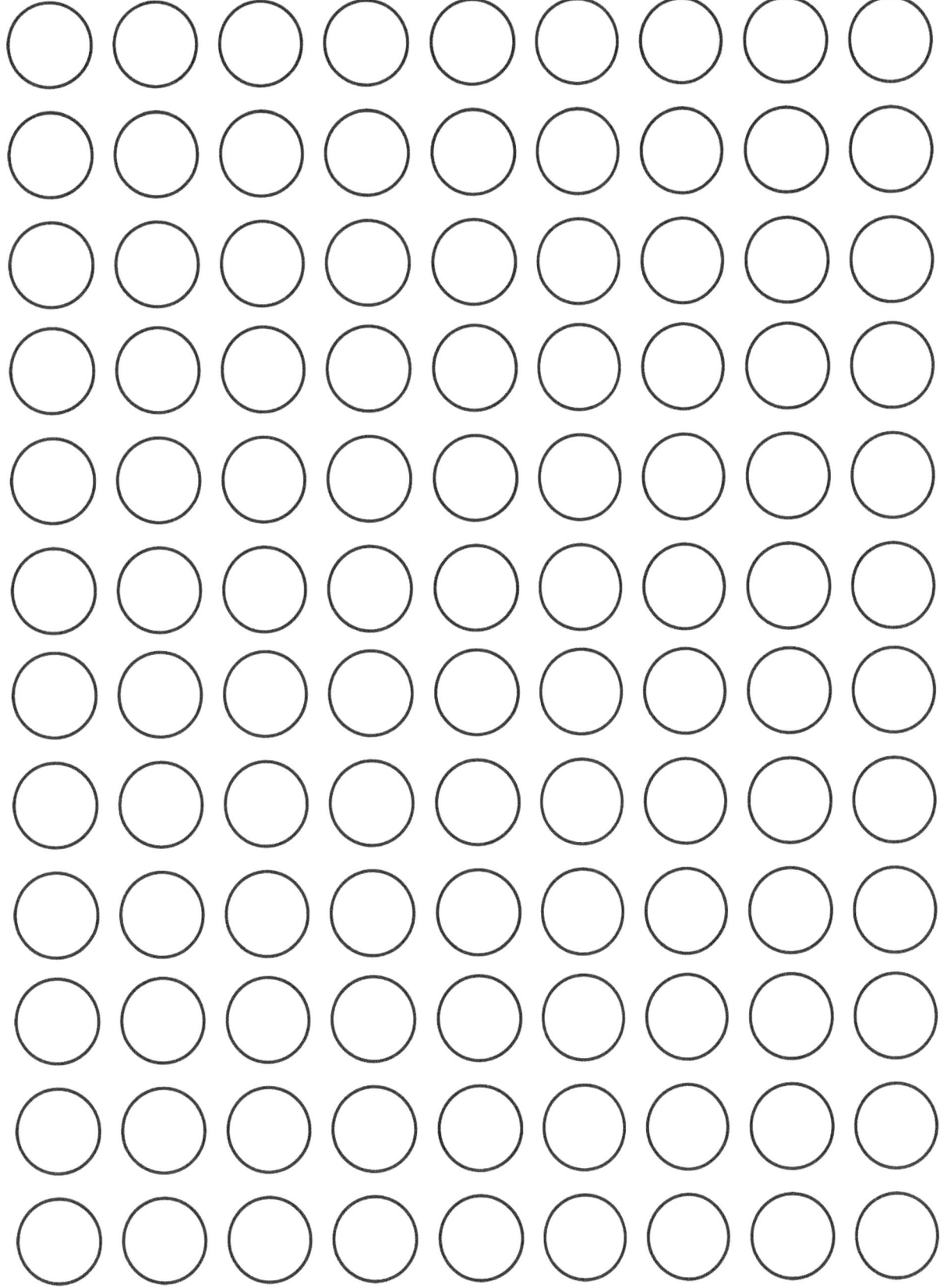

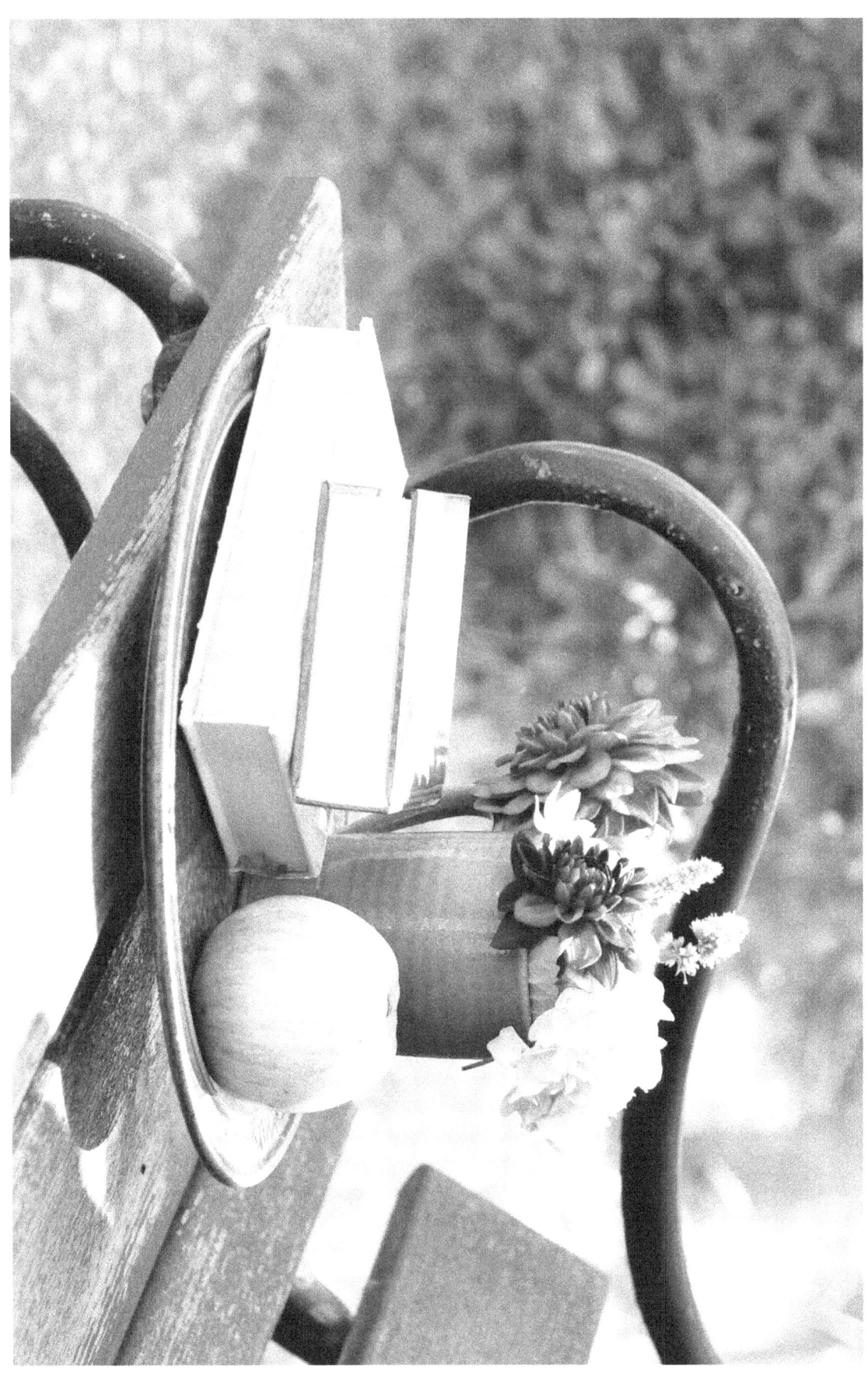

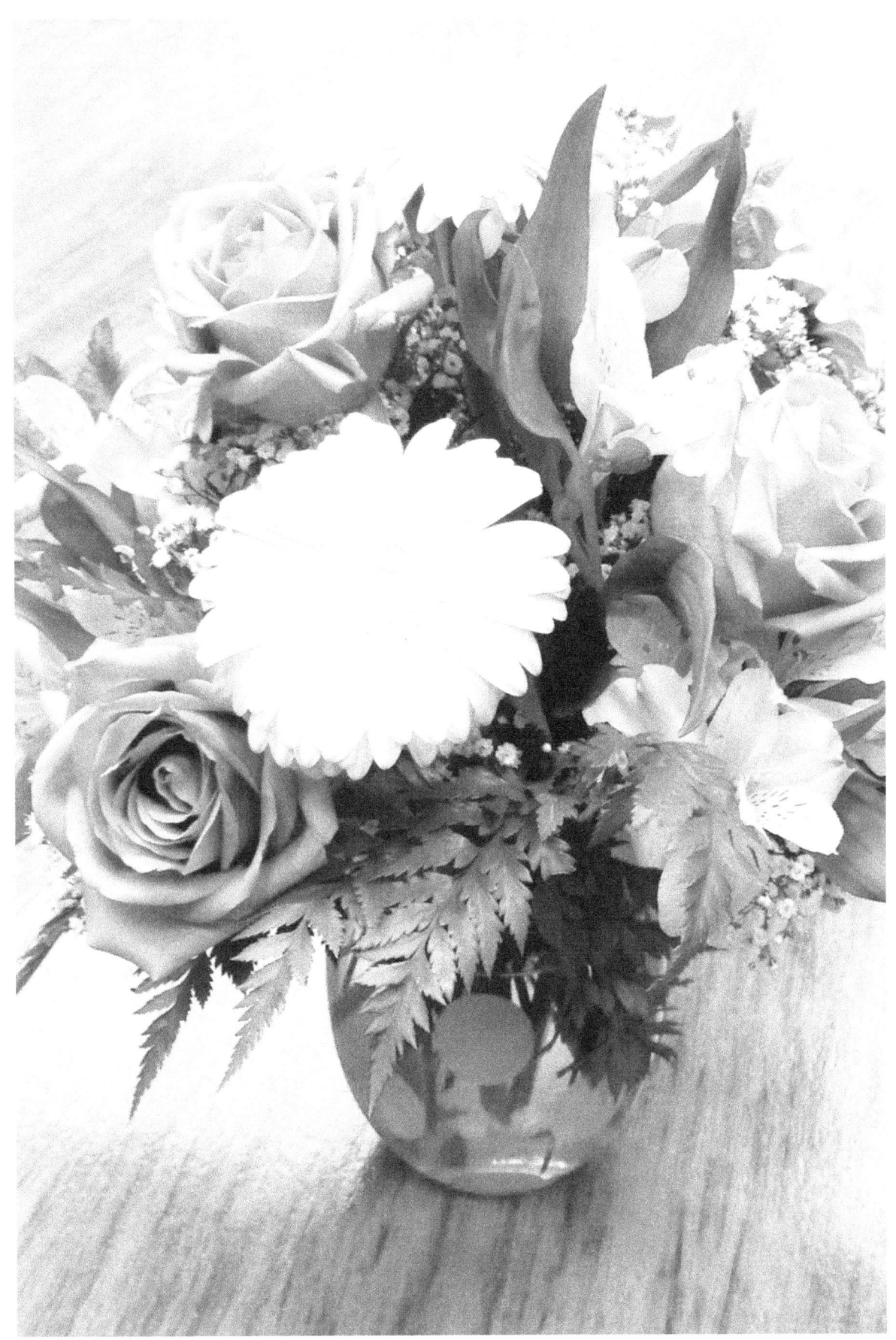

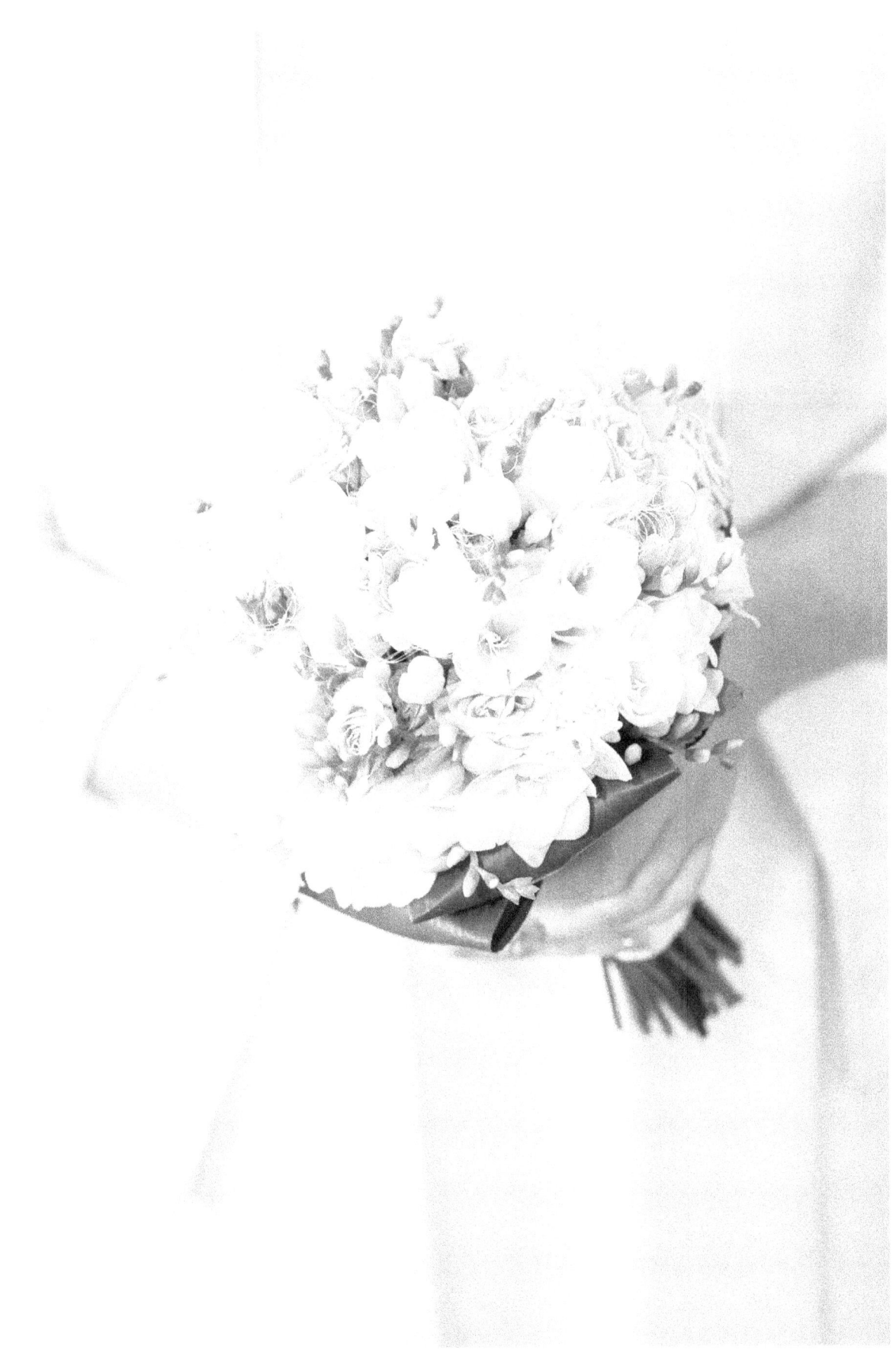

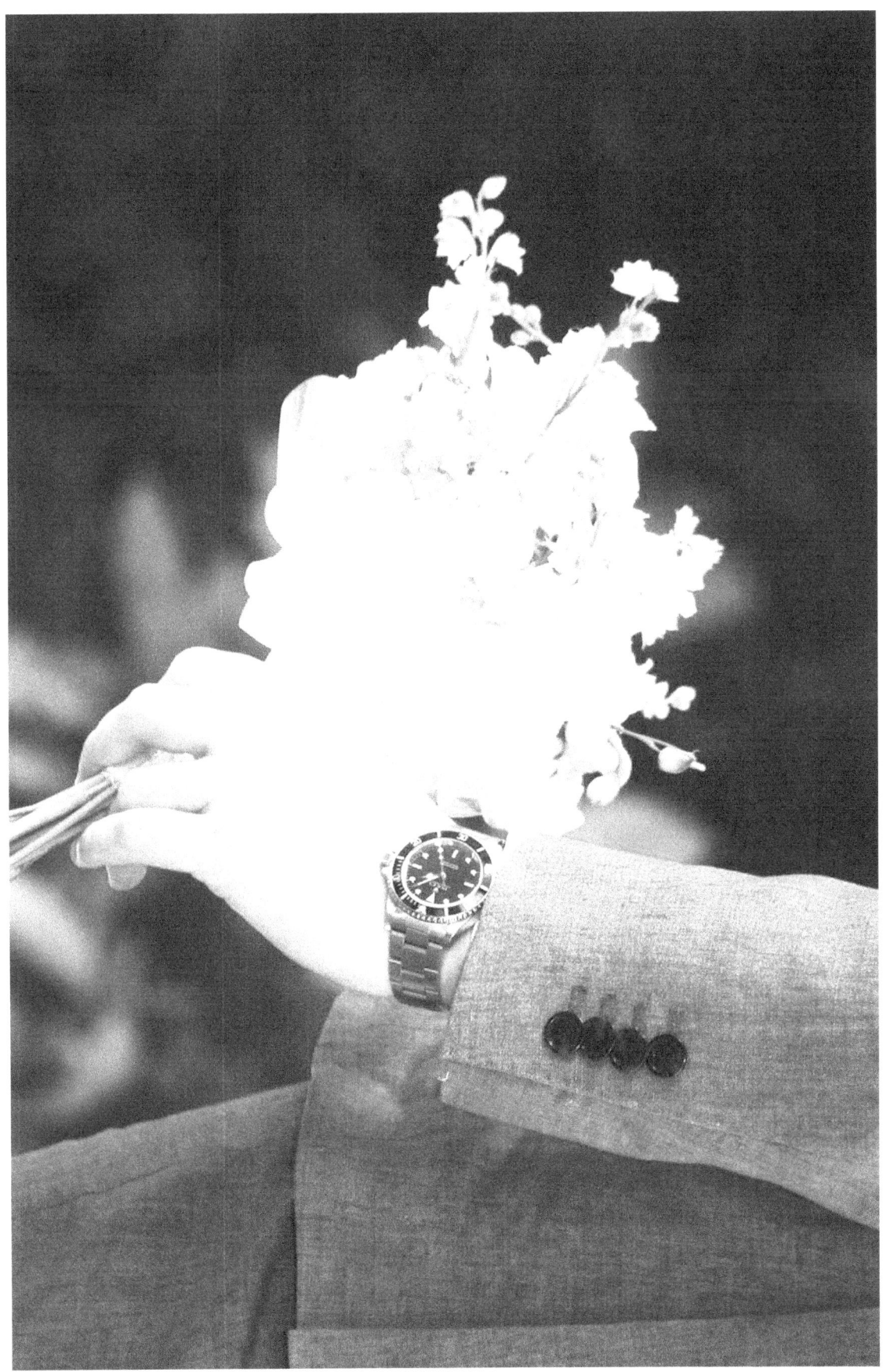

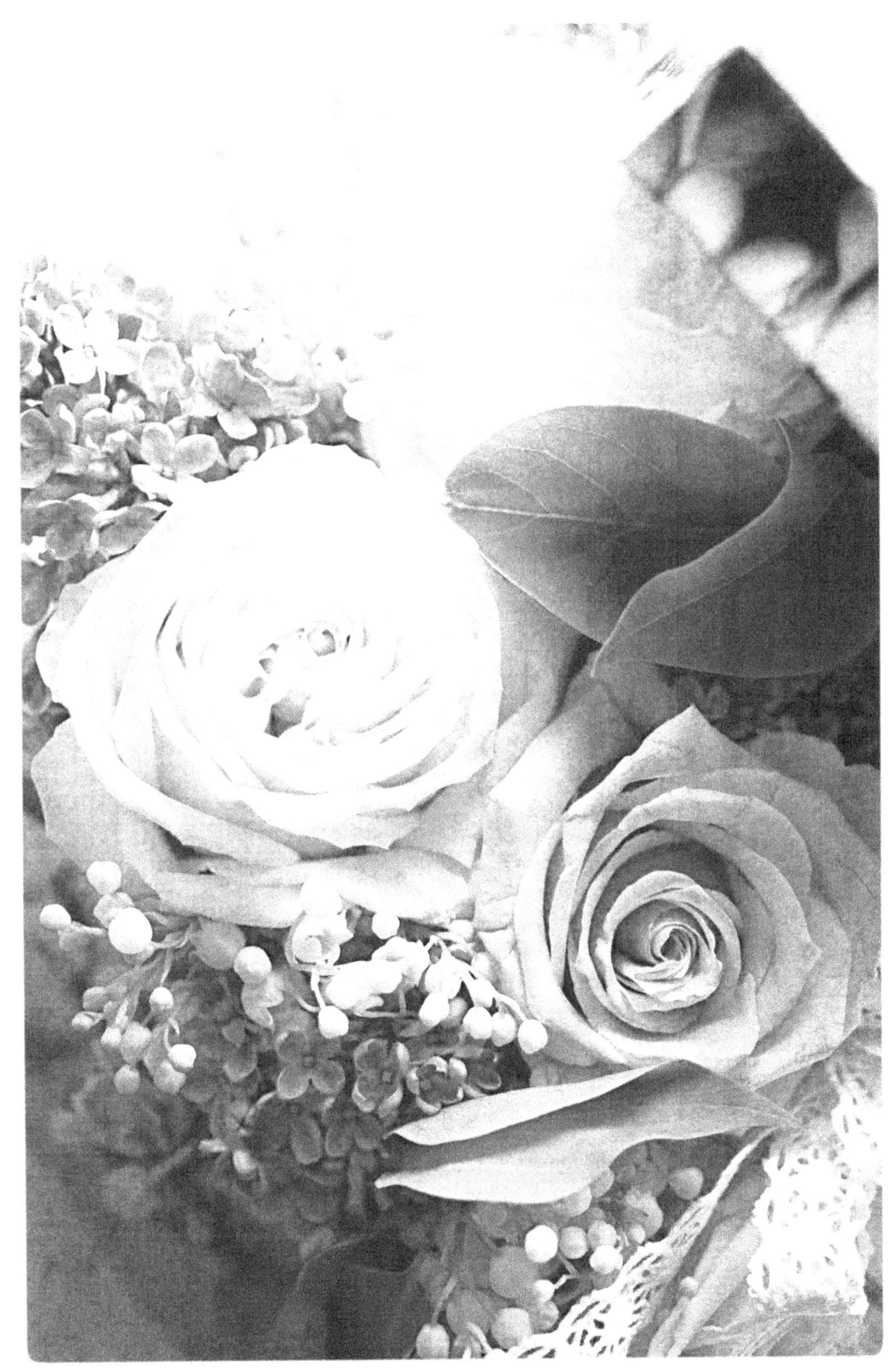

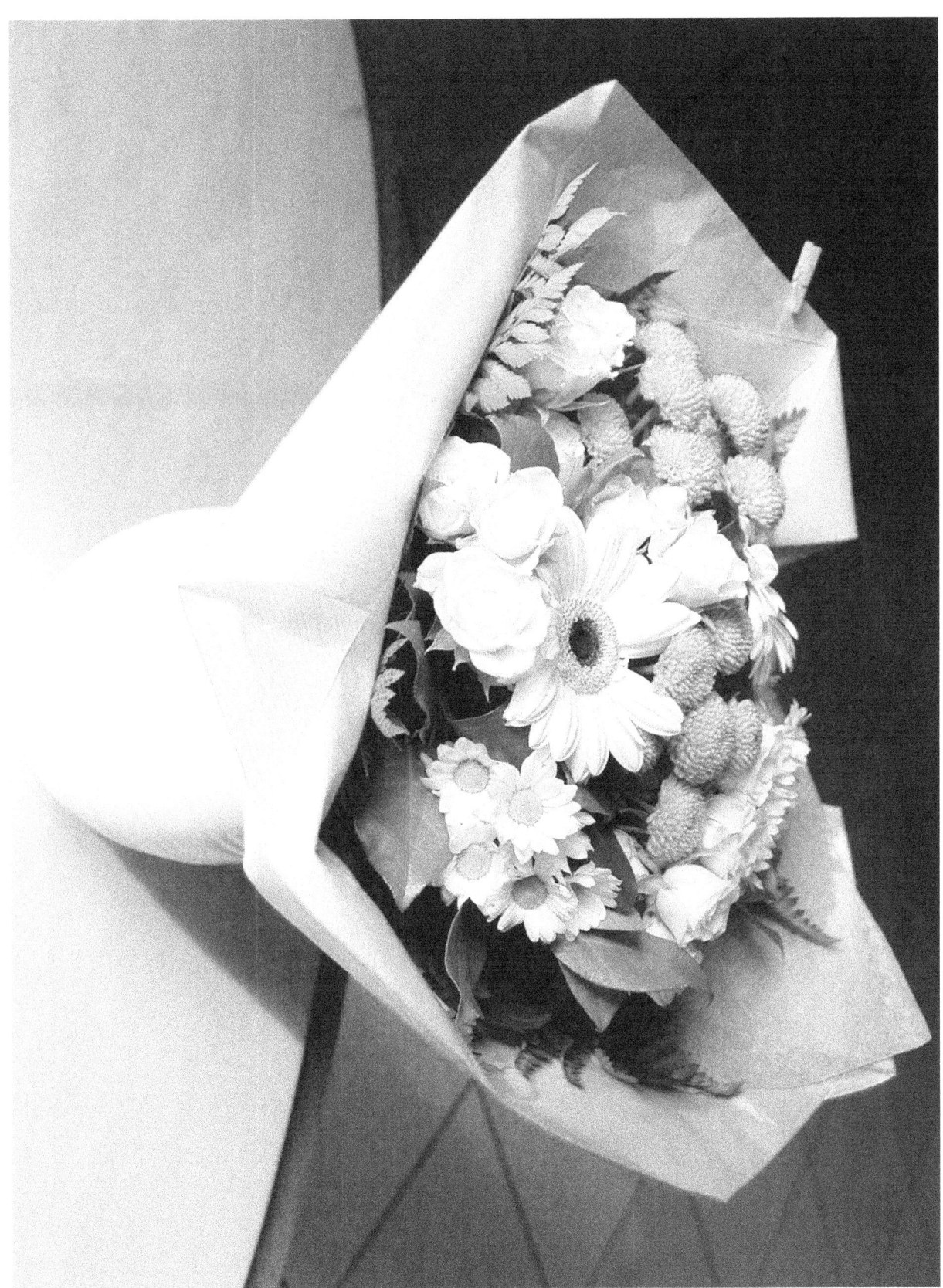

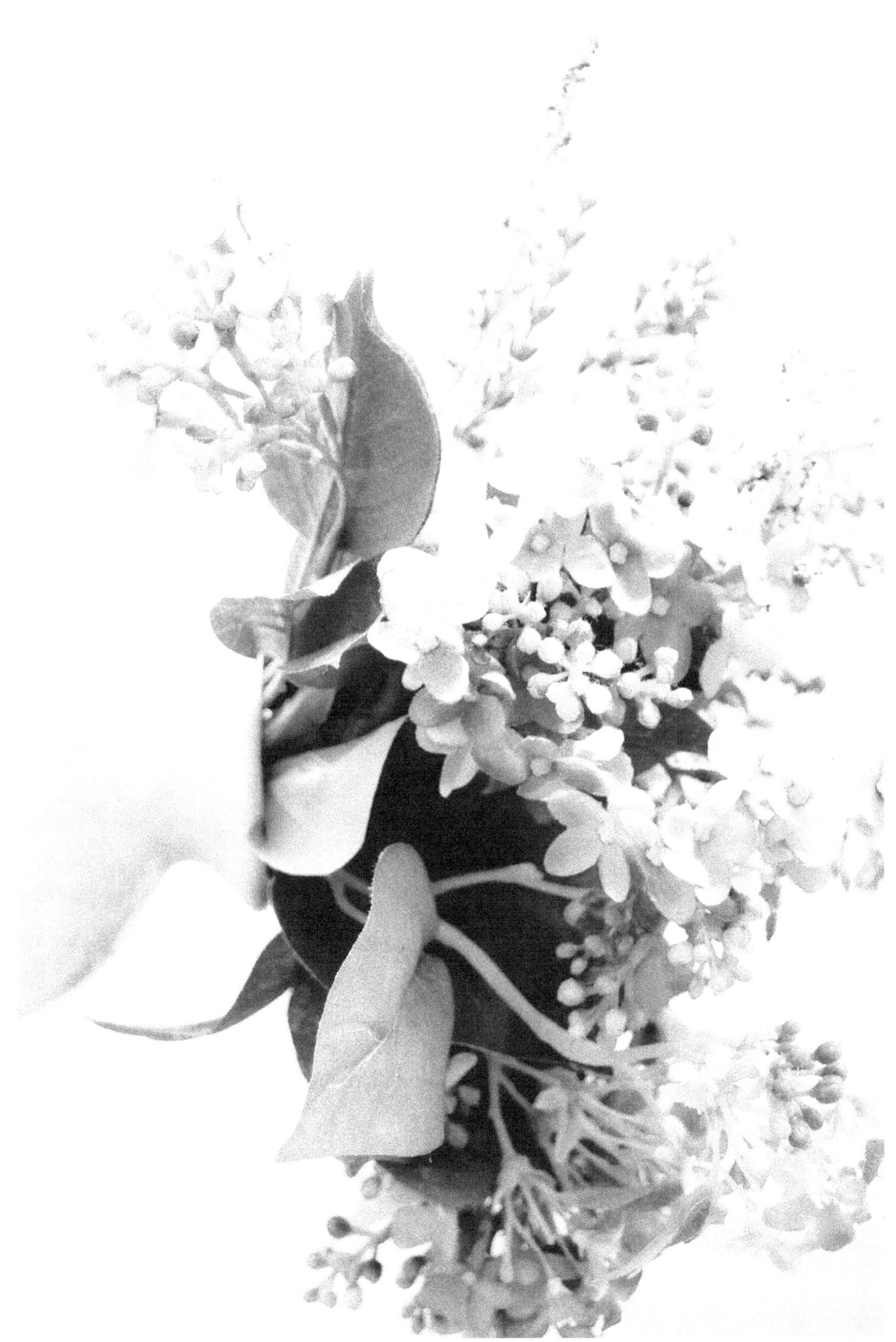

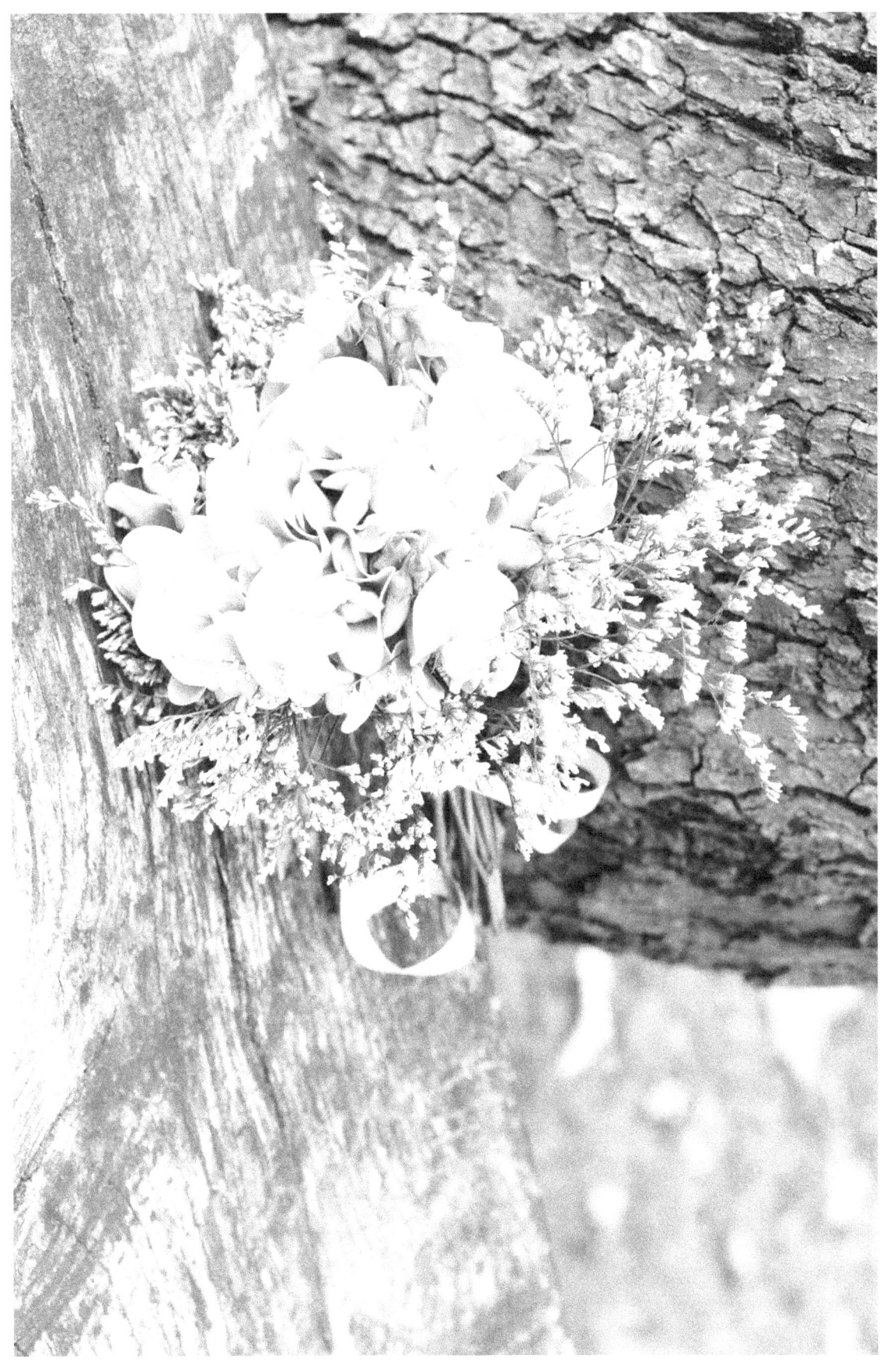

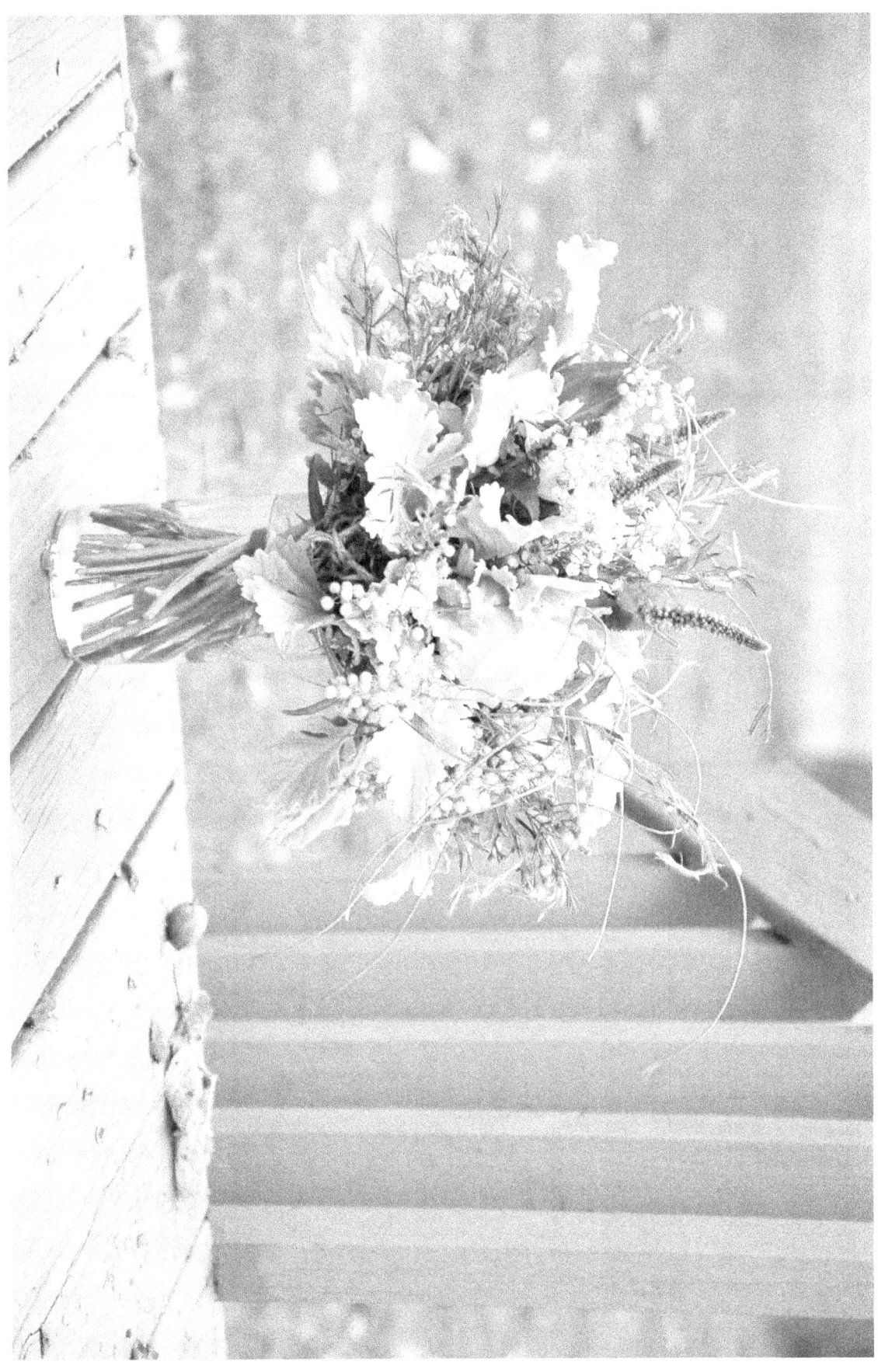

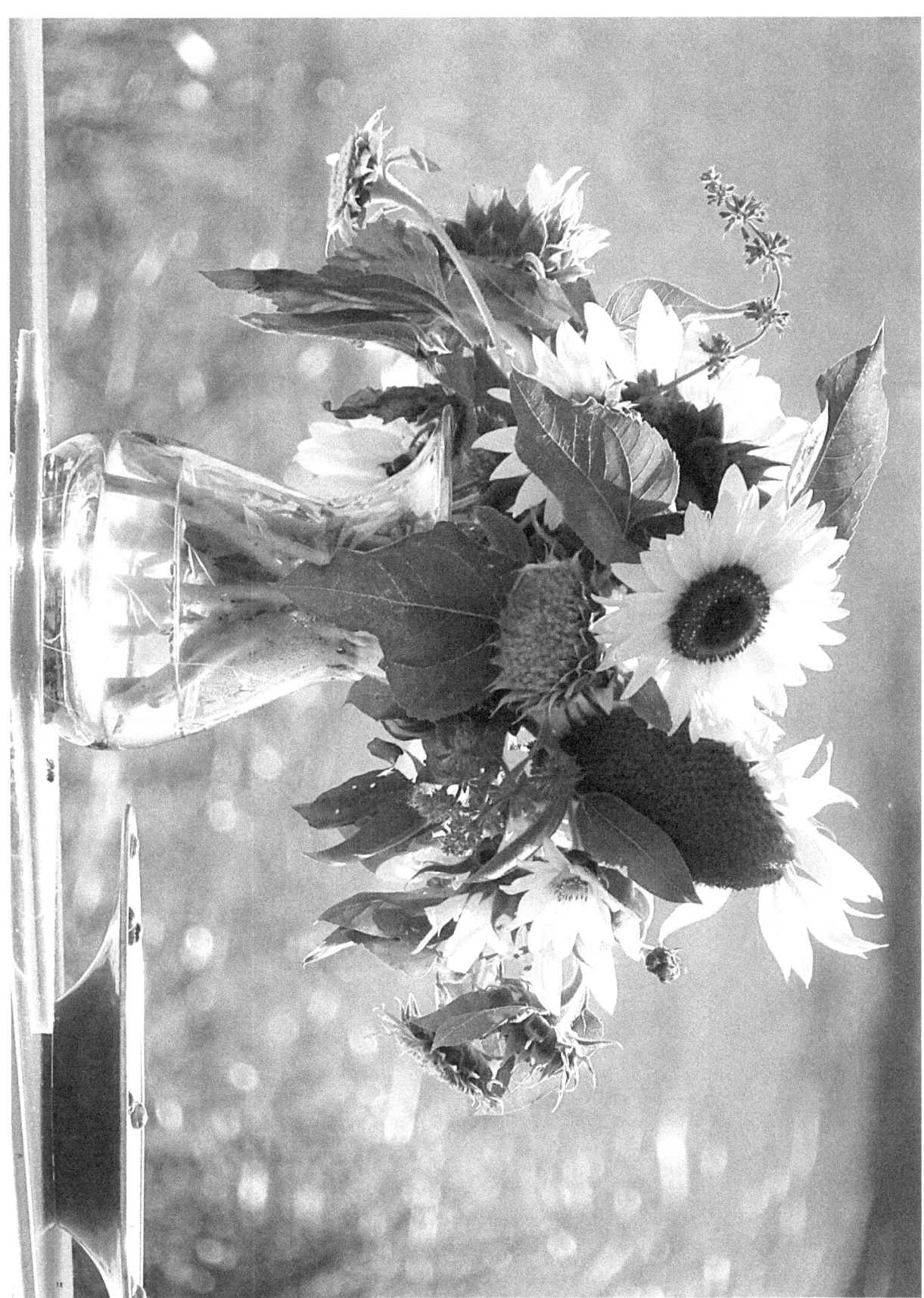

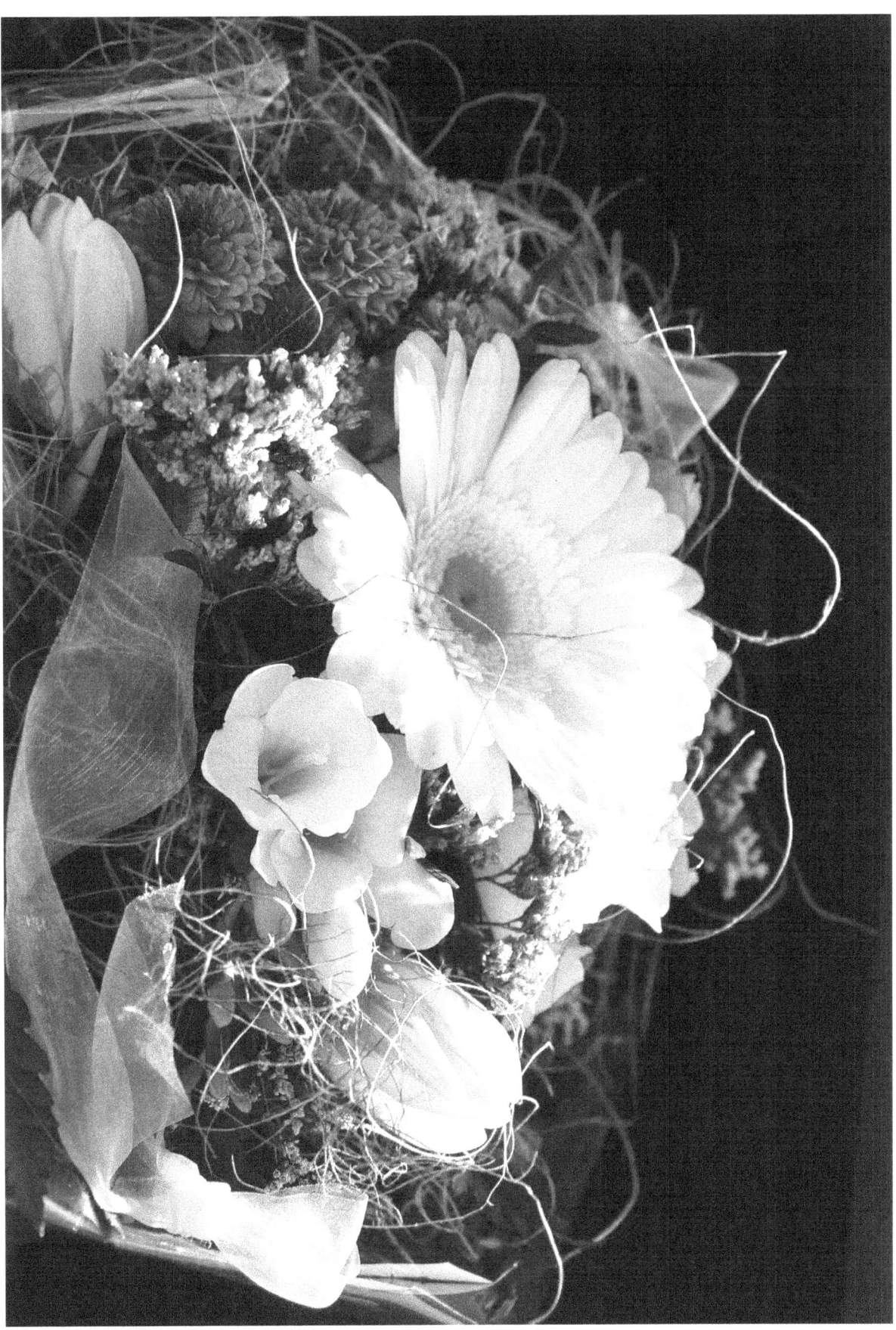

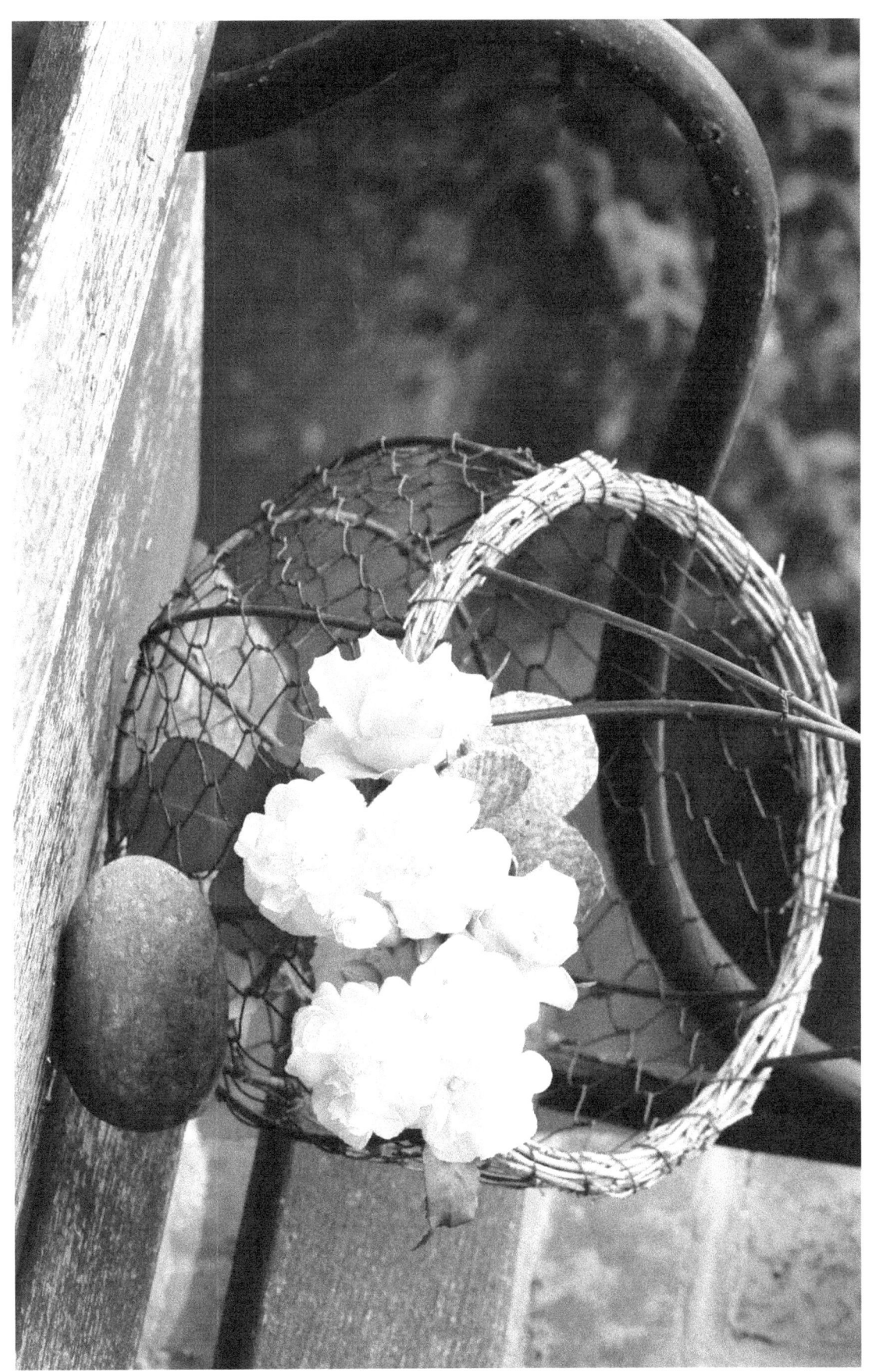

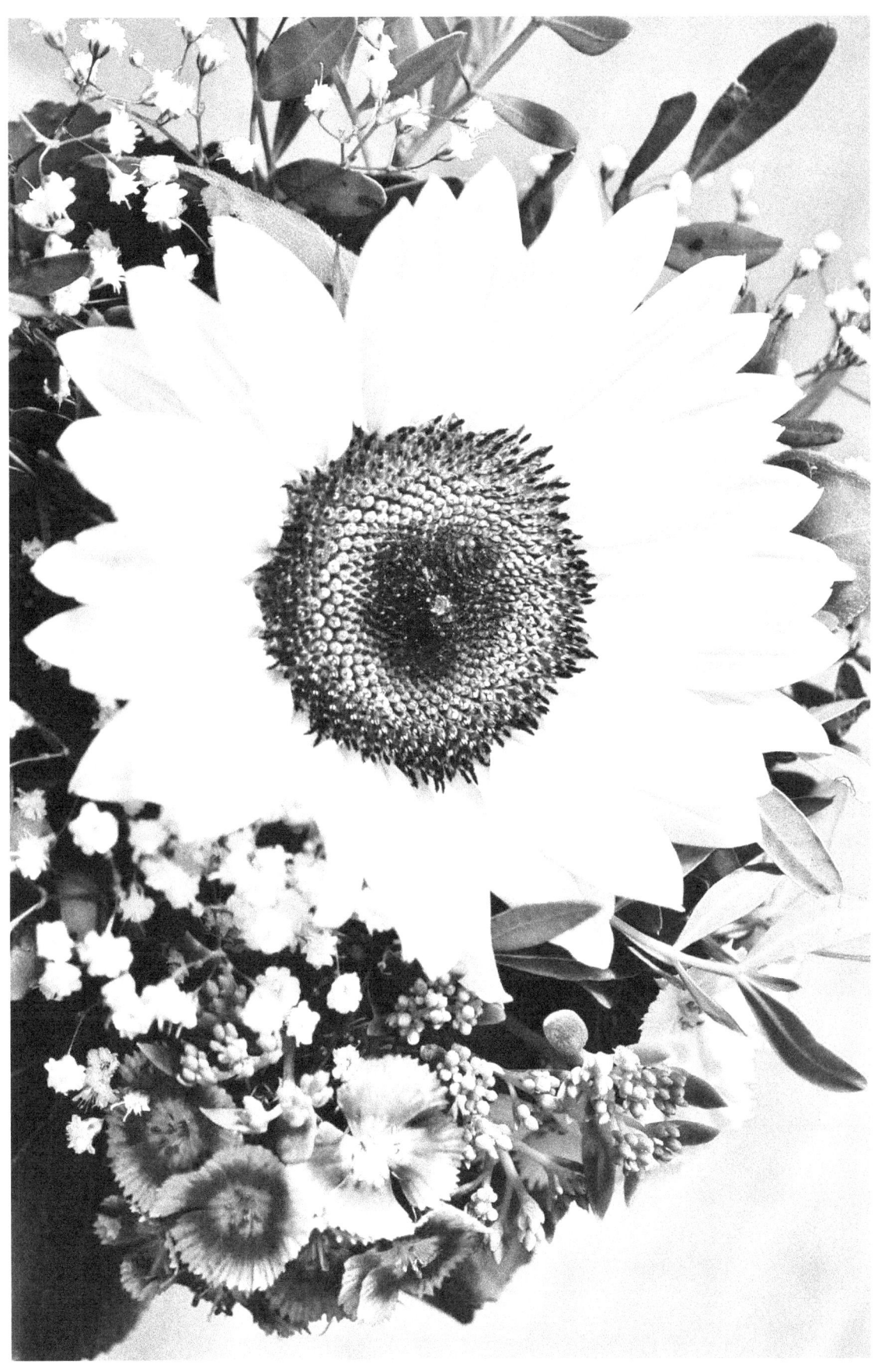

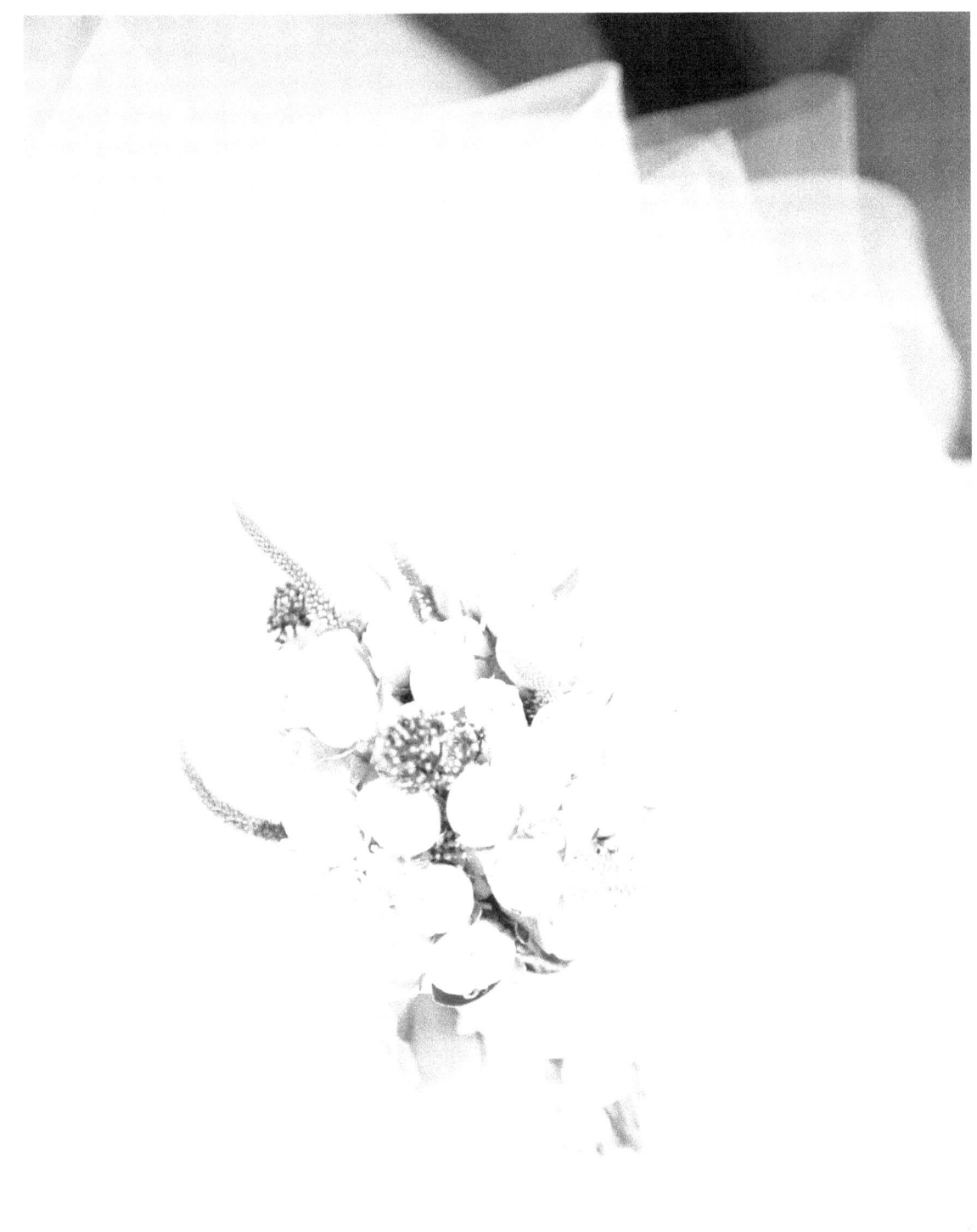

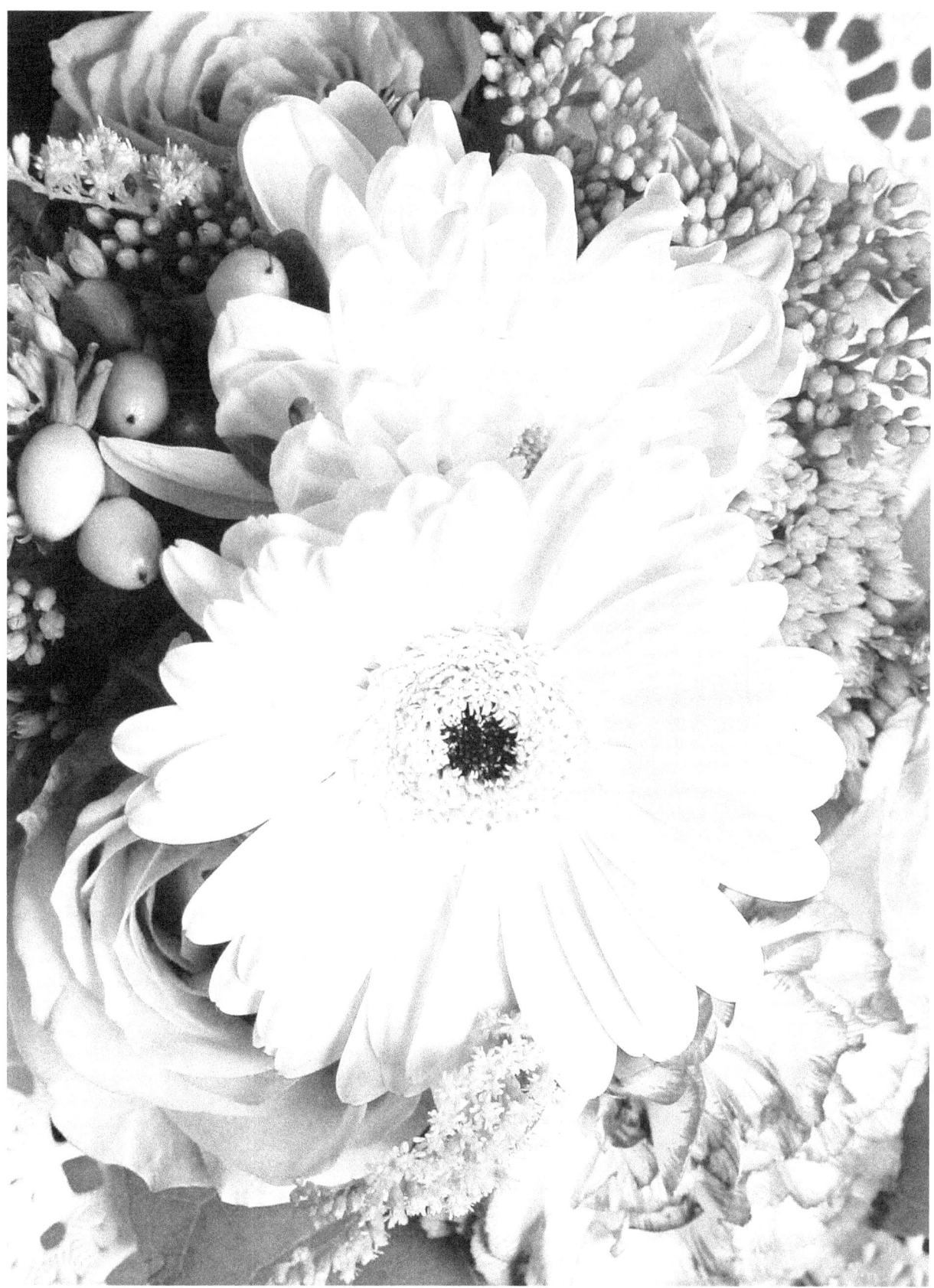

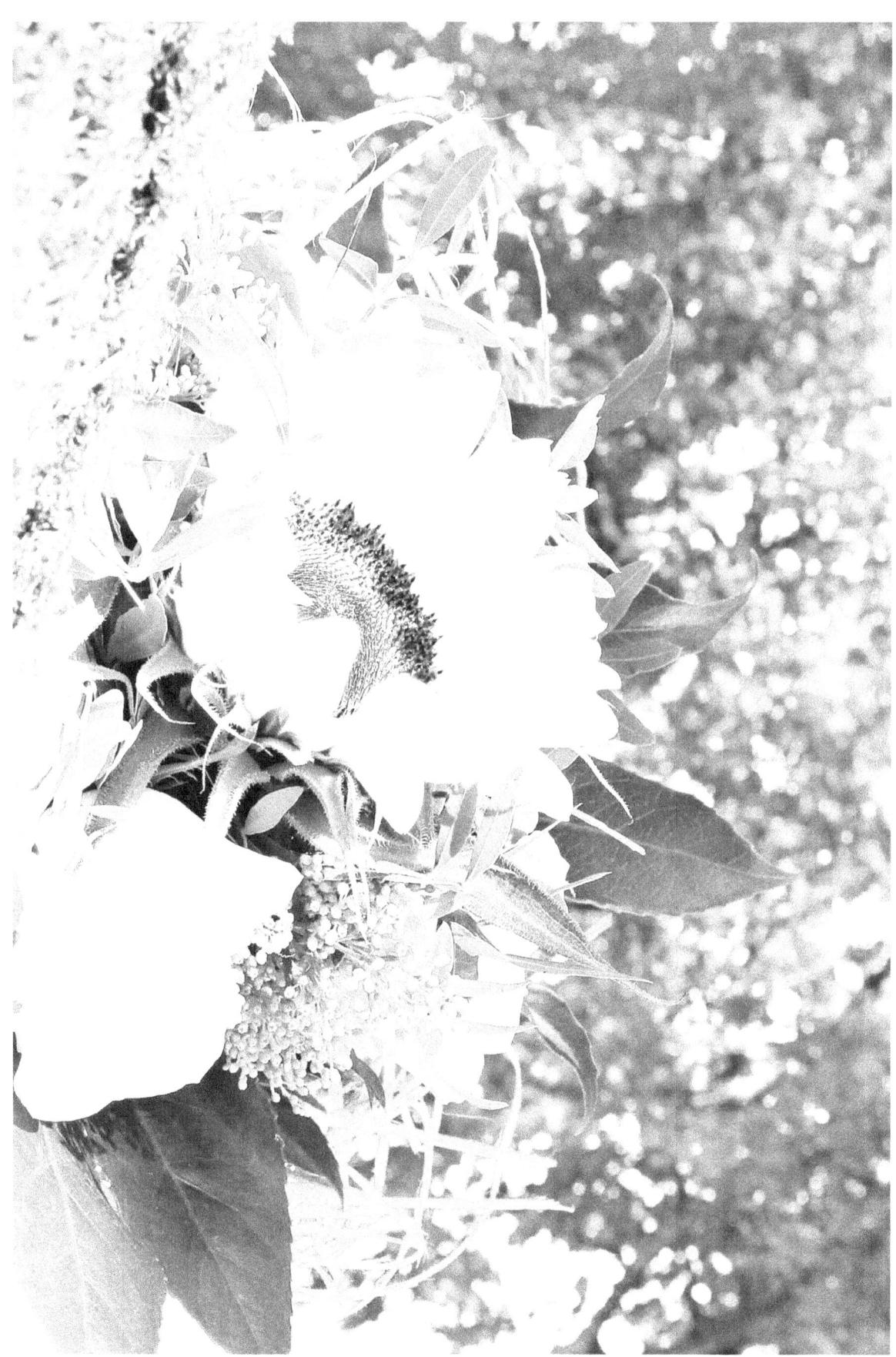

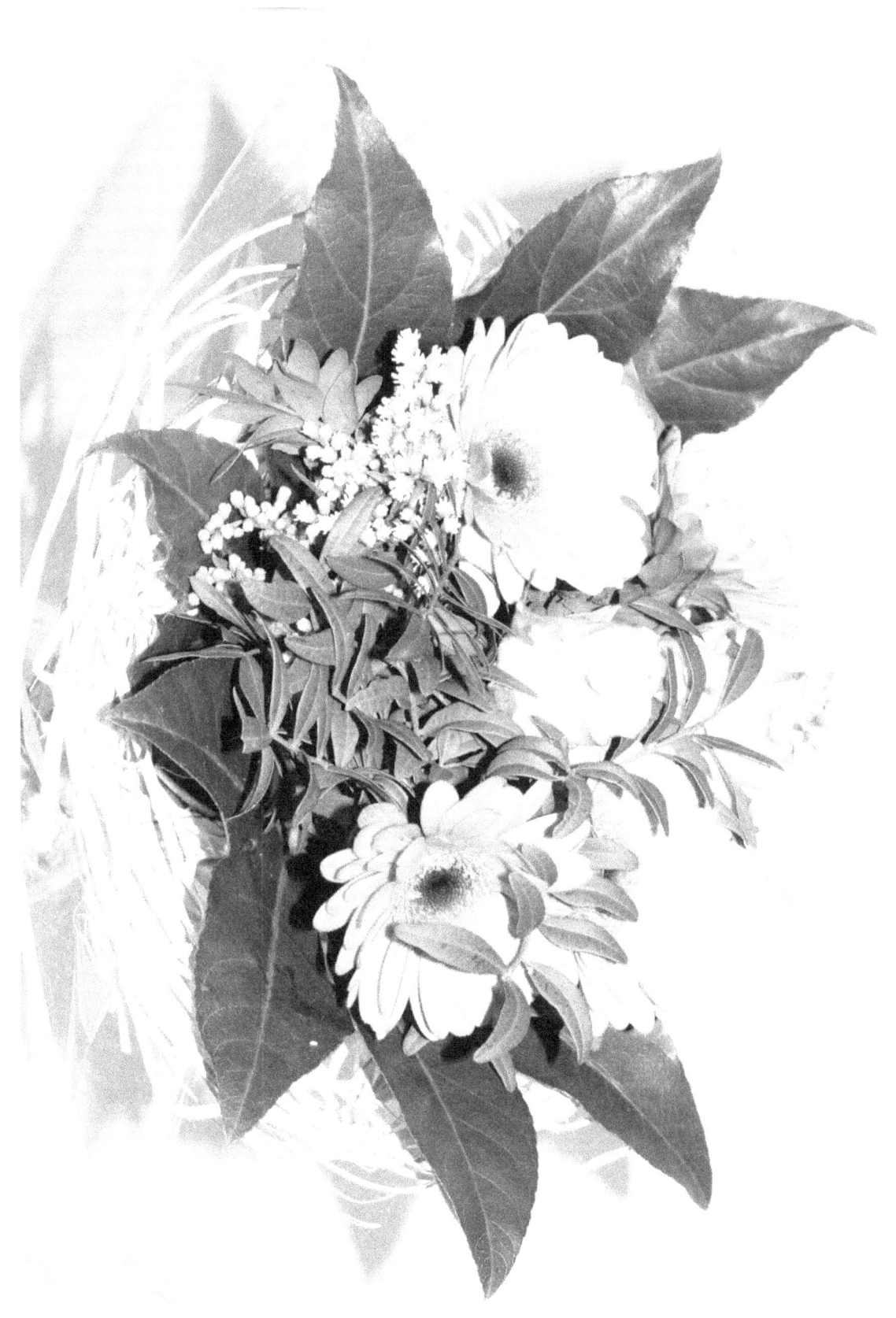

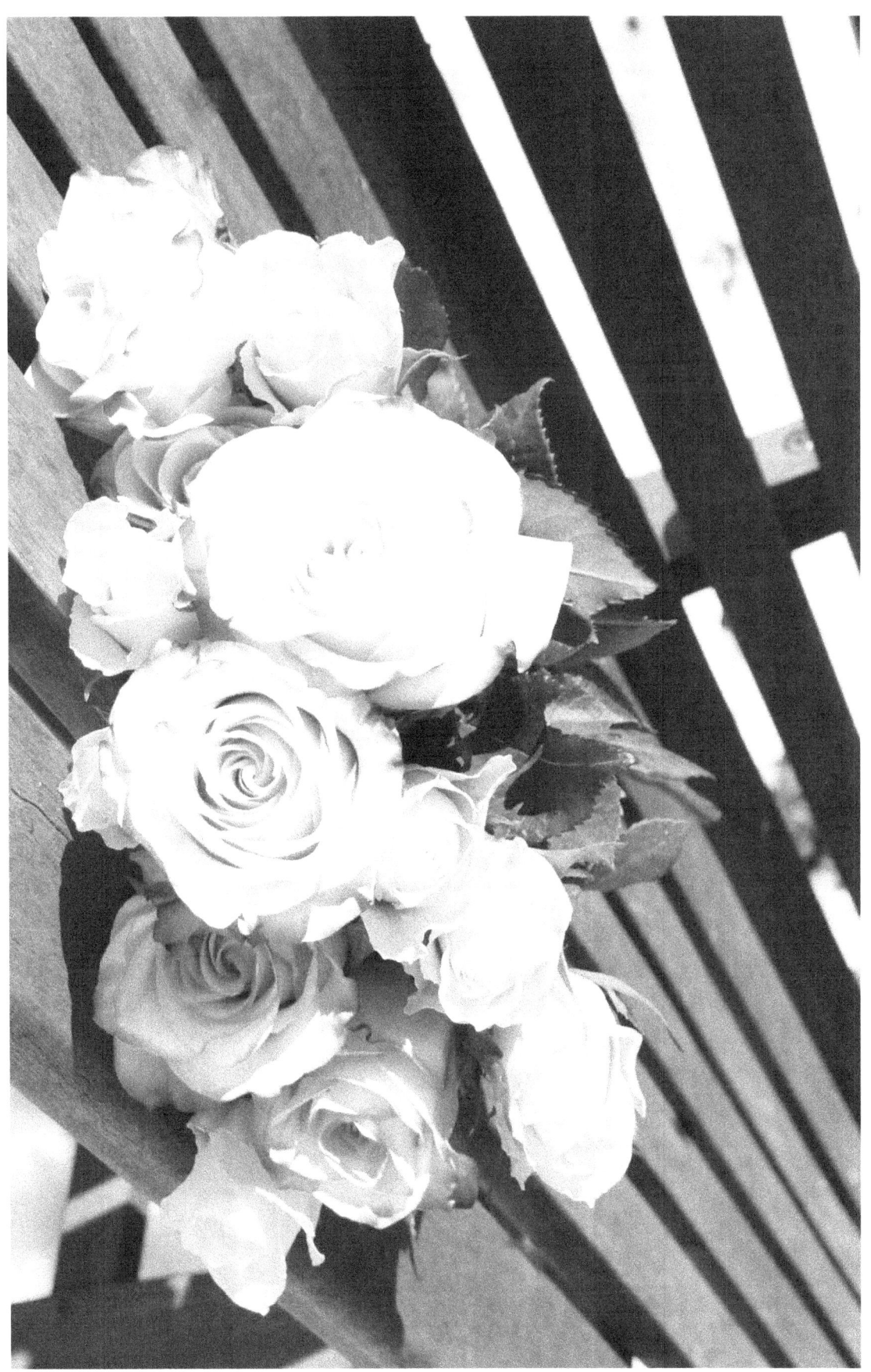

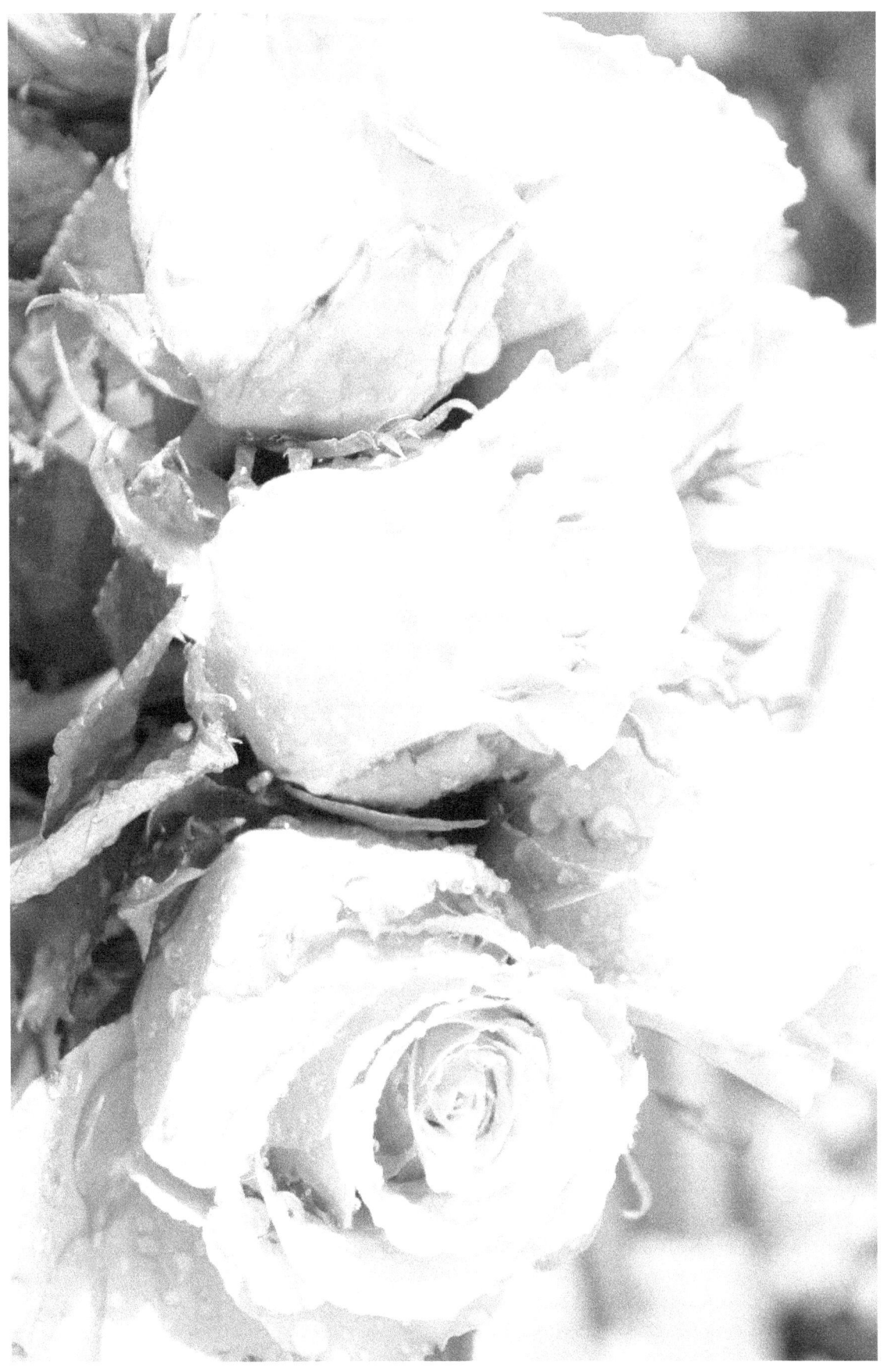

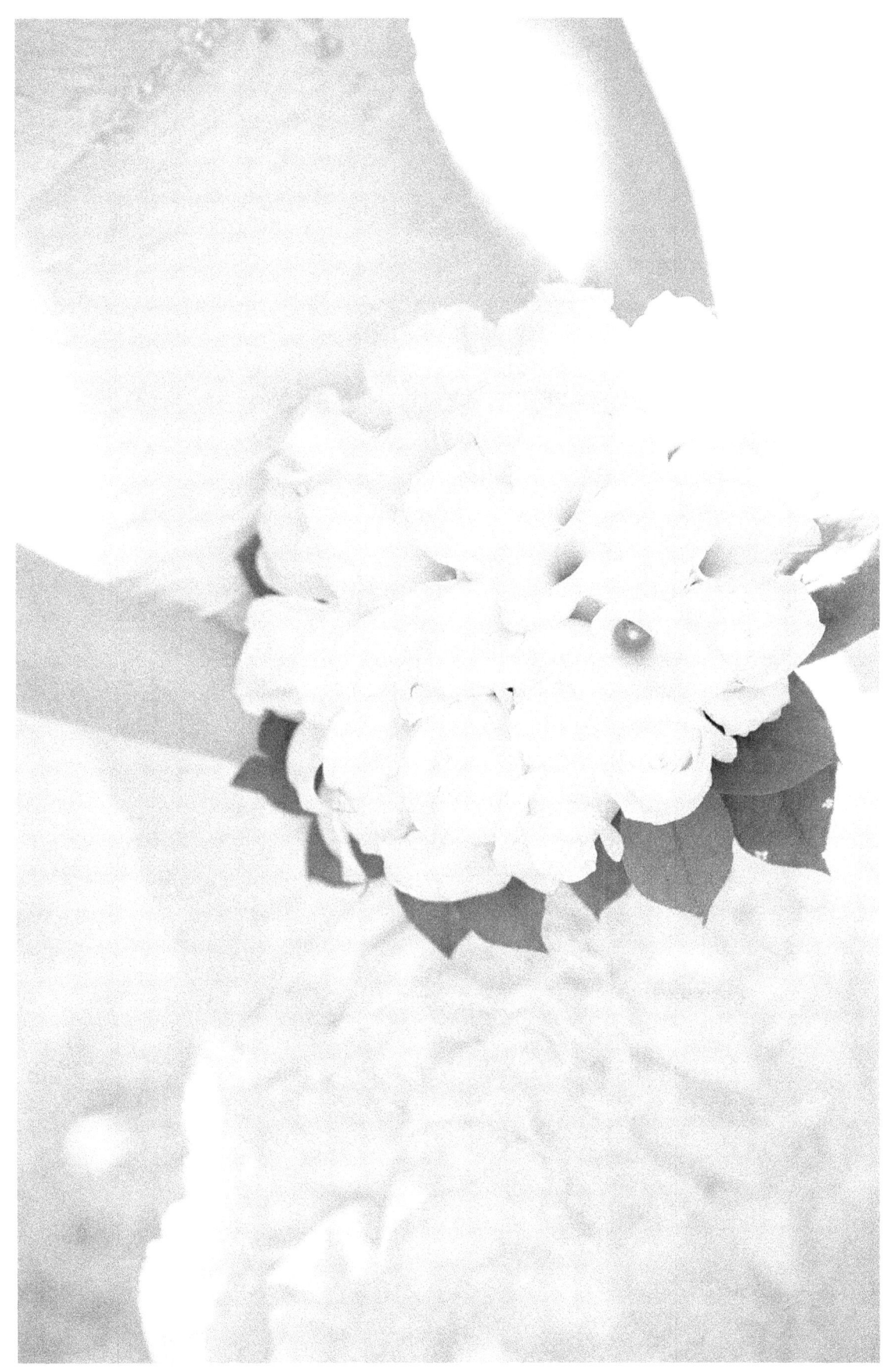

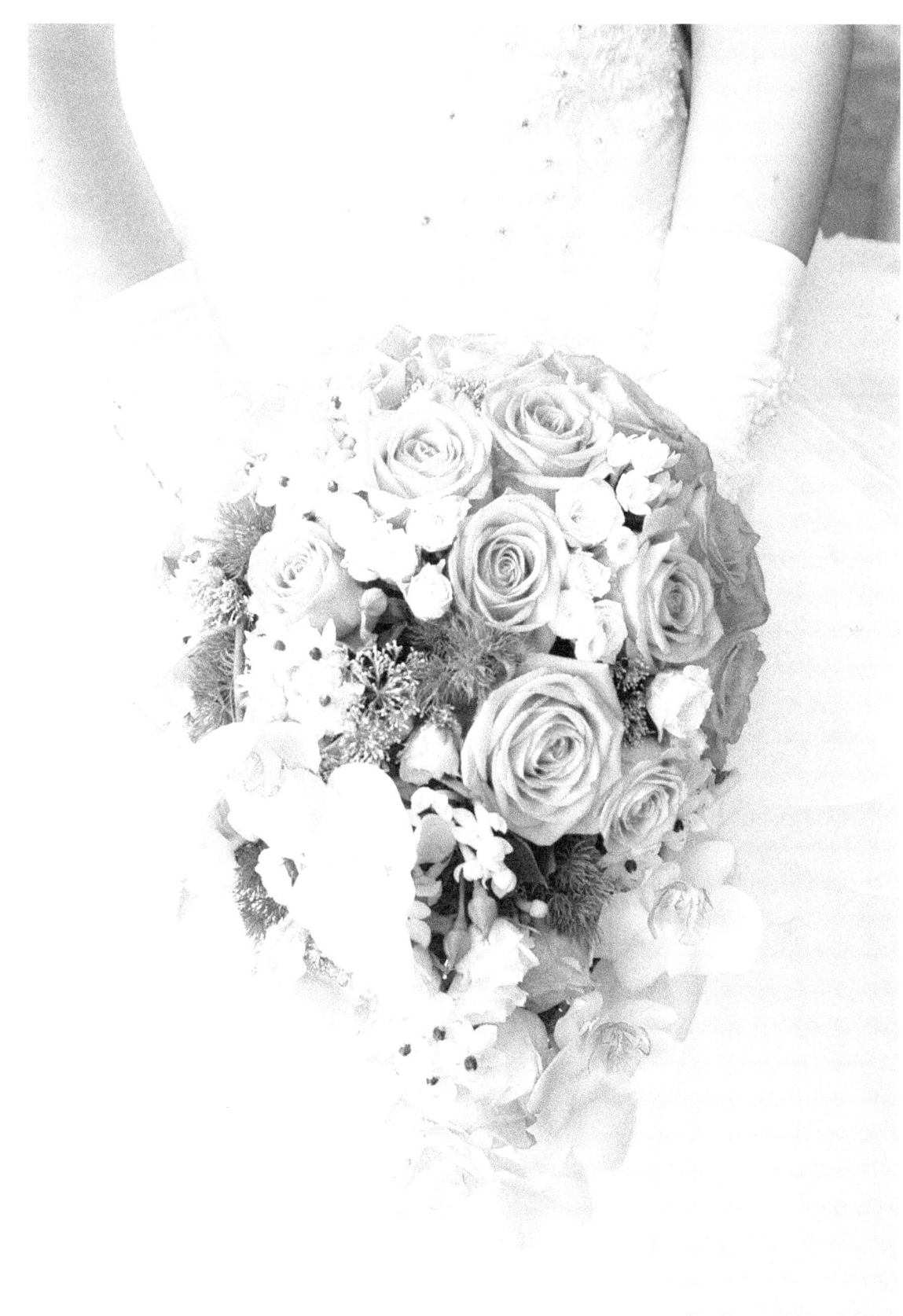

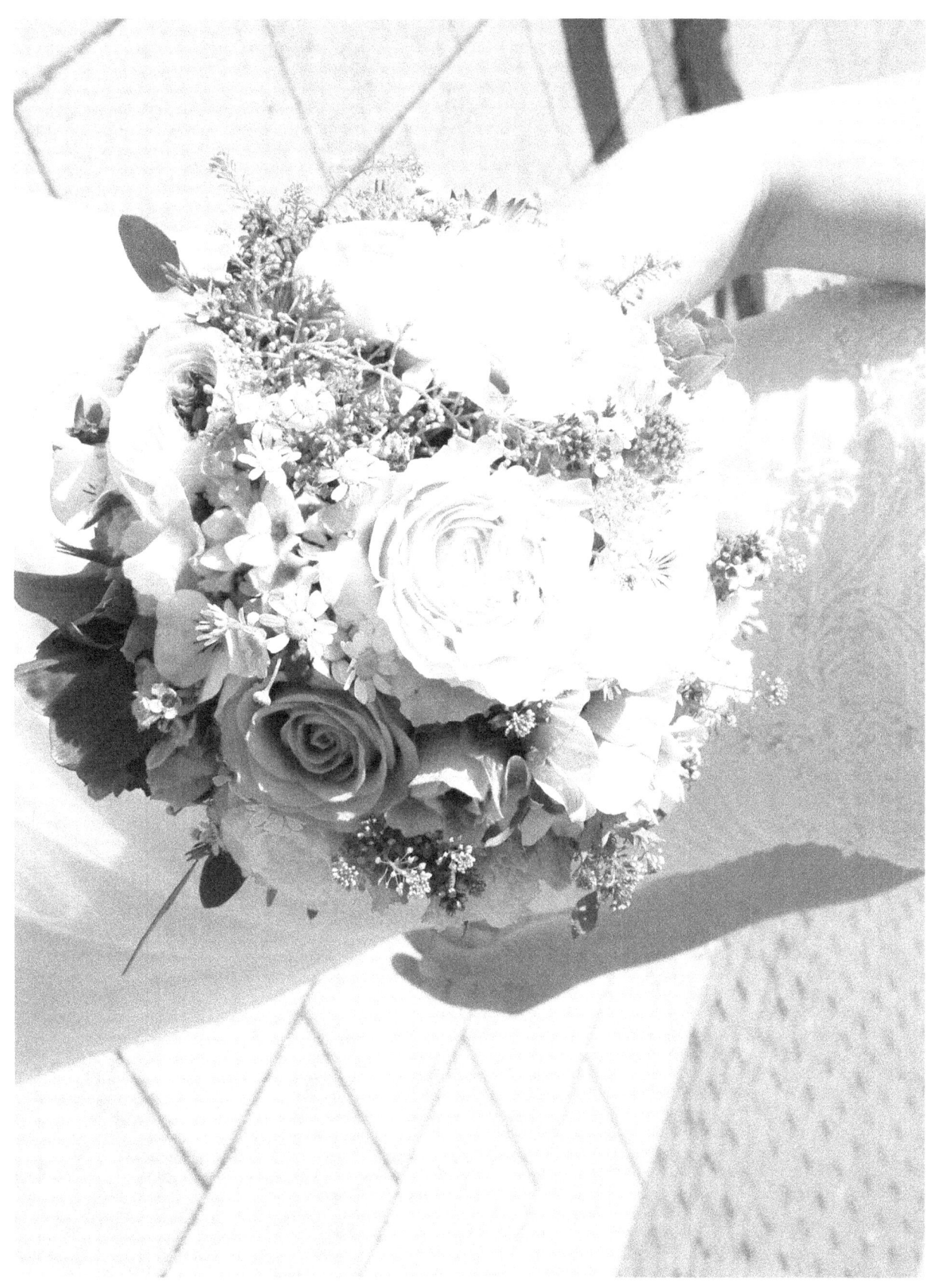

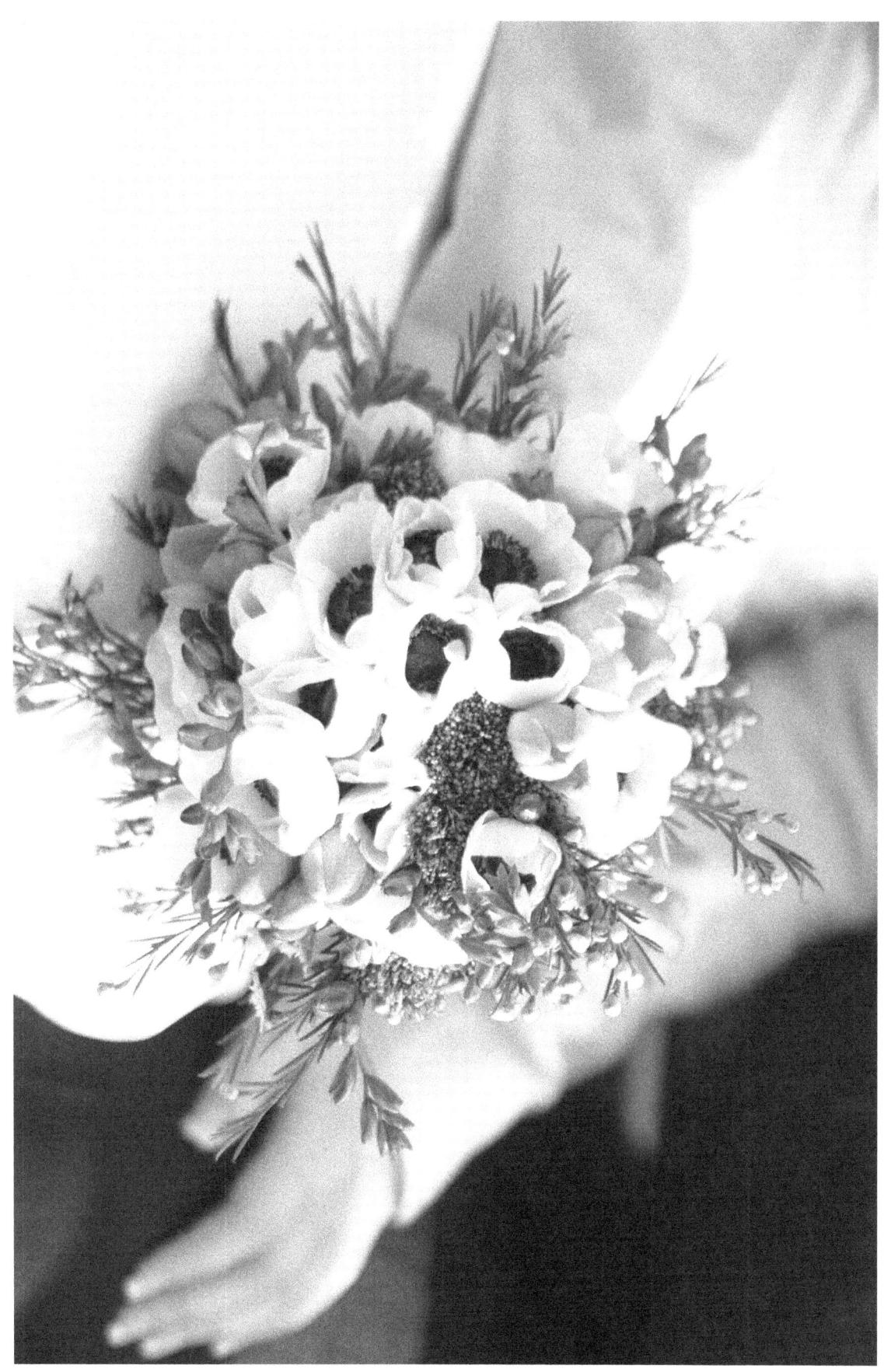

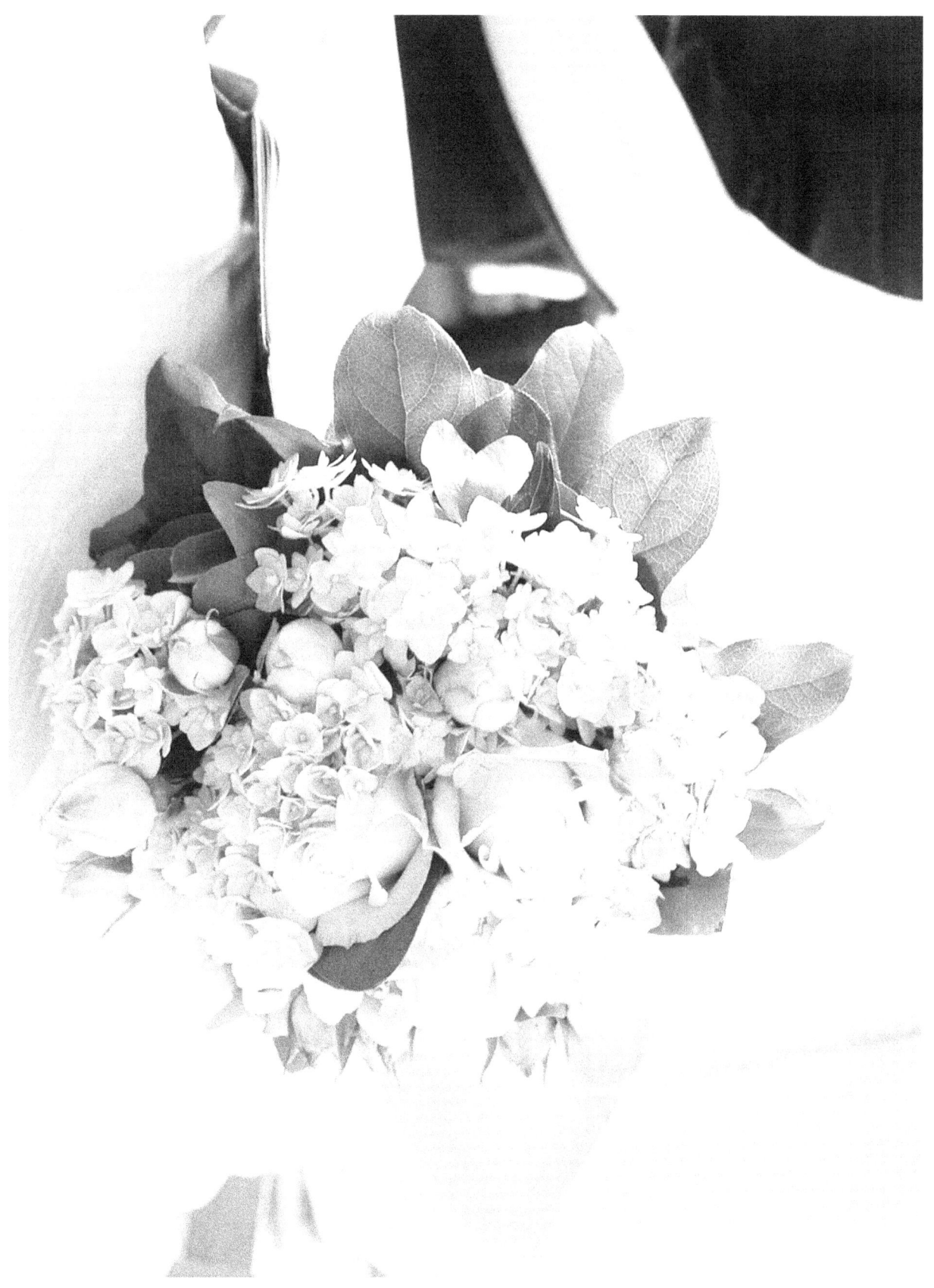

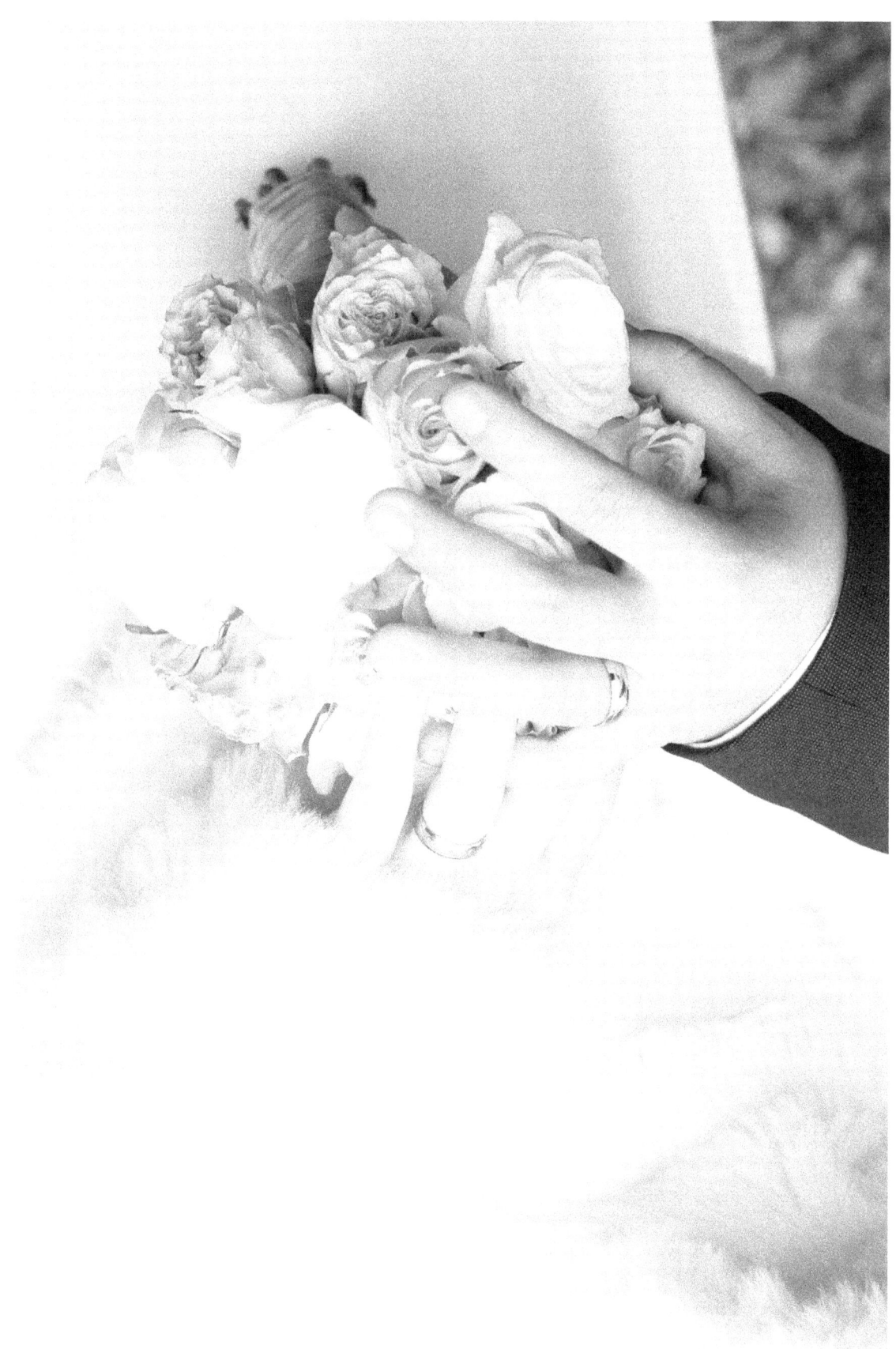

www.ingramcontent.com/pod-product-compliance
Lightning Source LLC
Chambersburg PA
CBHW080550190526
45169CB00007B/2716